José Roberto Cunha Lima
MAXMILIANO Sousa Veras

Moringa oleifera Lam.: medicinal applications and nutritional composition

José Roberto Cunha Lima
MAXMILIANO Sousa Veras

Moringa oleifera Lam.: medicinal applications and nutritional composition

Bromatological Analysis of Bioenergized Moringa oleifera Lamarck Leaves Grown in Parnaíba-PI

ScienciaScripts

Imprint

Any brand names and product names mentioned in this book are subject to trademark, brand or patent protection and are trademarks or registered trademarks of their respective holders. The use of brand names, product names, common names, trade names, product descriptions etc. even without a particular marking in this work is in no way to be construed to mean that such names may be regarded as unrestricted in respect of trademark and brand protection legislation and could thus be used by anyone.

Cover image: www.ingimage.com

This book is a translation from the original published under ISBN 978-620-2-19327-6.

Publisher:
Sciencia Scripts
is a trademark of
Dodo Books Indian Ocean Ltd. and OmniScriptum S.R.L publishing group

120 High Road, East Finchley, London, N2 9ED, United Kingdom
Str. Armeneasca 28/1, office 1, Chisinau MD-2012, Republic of Moldova, Europe
Printed at: see last page
ISBN: 978-620-7-27597-7

SUMMARY

I dedicate this to all my family members for their immense love and constant support, especially my mother Ana Gélia de Souza, who gave part of her life for my growth and character. To my father Paulo Luis Gomes Veras, who inspired me to value intellect and simplicity. To my sister Melissa de Souza Veras for her love and partnership. To my grandmothers Maria das Graças de Sousa, for her great love and motherly care, and Maria José Ferreira Sousa, for her kind and humble heart. To the late Jacy Gomes Veras and Pedro Alves Marinho; grandparents who gave me great affection and motivation in life.

To my advisor, Leiz Maria Costa Veras, for her partnership, encouragement and opportunities for academic growth during my degree; and to my co-supervisor, José Roberto da Cunha Lima, for his teachings and friendship during the course.

ACKNOWLEDGMENTS

I glorify and thank God first of all for the gift of life, for knowing him and being able to call him my son. Without his powerful hand on me, my future would be without direction. To him I give all honor, glory and power, for he alone is worthy. All my willpower, dedication and love for Nutritional Science have been given to me by him, so I give thanks to my God for his blessings during my four-year course. He is everything to me!

I am immensely grateful to my parents Ana Gélia and Paulo Luis for giving me unconditional love and affection from the beginning of my life. My motivation and will to succeed came from the difficulties we experienced, and with their strength, we managed to overcome them. All the teachings and times you took me to school as a child were indispensable. You are responsible for the man I have become and all my achievements. I love you all very much!

To my sister Melissa, for the love you have for me and your example of dedication and willpower that inspired me to be a better student. I love you too little sister!

To my grandmothers Maria das Graças and Maria José, for their love, advice and all the investments they made for my future. I have them as my second mother, and I'm proud to say that I'm a blessed grandson for having them in my life. You live in my heart!

To my dear advisor Leiz, for guiding me through this work, as well as believing in my potential and giving me close experience of research into chemistry and biochemistry, two areas I hold in high regard. I have enormous admiration and affection for you, my master. Thank you for everything!

To my co-supervisor José Roberto, for his teachings and experiences throughout my degree, and for contributing directly to the analysis by always being willing to teach and listen. I admire you for the professional and human being you are, so thank you, Master!

To Professor Edson Teófilo, for accepting my research proposal and donating the moringa produced by his company for the benefit of science and people's quality of life. I am immensely grateful!

"But let him who boasts, boast in this, that he knows me, and knows that I, the Lord, do mercy and justice on earth, for in these things I delight, says the Lord."

(Jeremiah 9:24)

SUMMARY

In contemporary times, the context of healthy eating occupies a great deal of space, especially in the social media, and is propagated as a lifestyle to be adopted. Eating out has grown in recent years due to a demand for practicality, although most meals are still eaten at home. From this point of view, the literature has focused on *Moringa oleifera Lam.* due to its nutritional and pharmacological potential. *Moringa oleifera* Lamarck or *Moringa pterygosperma* Gaertner, is a leguminous plant in the Moringaceae family, popularly known in Brazil as moringa, white lily or okra. *M. oleifera* contains essential nutrients such as vitamins, minerals, amino acids, antioxidant and anti-inflammatory compounds, as well as omega 3 and 6 fatty acids. The consumption of moringa leaves is a strategy that can guarantee prevention and protection against various diseases, as well as boosting the body's intake of essential nutrients. In view of the above, the work has a quantitative, qualitative, exploratory, descriptive, experimental, documentary, pure and bibliographic approach. The pH, moisture, ash, protein, lipid, carbohydrate and calorie values of the bioenergized moringa leaves were determined. The following values were obtained from the analysis: pH (5.270), humidity (33.810 %), ash (8.566 g), protein (9.028 g), lipids (8.207 g), carbohydrates (40.380 g) and calories (271.506 Kcal). The nutritional potential of the leaves studied was confirmed in the light of the bibliographic data already recommended through the similarities between most of the analyses. Bioenergization therefore plays a fundamental role in improving the nutritional value of vegetables.

Keywords: *Moringa oleifera* Lam.; Bioenergization; Bromatological analysis; Nutritional characterization.

1 INTRODUCTIONÇTHE

In contemporary times, the context of healthy eating occupies a great deal of space, especially in the social media, and is propagated as a lifestyle to be adopted. On the other hand, obesity and cardiovascular diseases have been worrying the population and especially health professionals, as they are appearing on a large scale and can be justified by a lack of adequate nutrition.

Changes in eating habits over the years have been observed through the high intake of high-calorie foods and the low consumption of fruit and vegetables (MORAIS et al., 2016). The strategy proposed by the World Health Organization (WHO) in 2003, entitled Global Strategy for Food, Physical Activity and Health, aims to reduce the consumption of high-calorie and high-sodium foods, low nutritional content, saturated and trans fats and refined carbohydrates (CLARO, et al., 2015).

Eating out has grown in recent years due to a demand for convenience, although most meals are still eaten at home. As a result, there has been an increase in the number of establishments in the food sector, especially *fast-food* chains, restaurants and other food service outlets that mostly offer food with low nutritional value (BEZERRA, et al., 2017). Given these circumstances, a food or product that offers practicality and reduced cost could help to improve nutritional intake and prevent the emergence of chronic non-communicable diseases (CNCDs).

From this point of view, the literature has focused on *Moringa oleifera Lam.* due to its nutritional and pharmacological potential. The tree has been the subject of research over the years, one of the objectives being its application in the fight against malnutrition, since its leaves can offer adequate nutrients at a low cost. They could possibly be used as a food supplement for children in the first 1000 days of life in order to intervene in chronic malnutrition (LEONE et al., 2015).

In terms of macronutrients and other elements, it can be seen that the leaves have a good proportion of proteins containing all the essential amino acids, lipids, fiber, high levels of β-carotene and lutein. In addition, there are few anti-nutritional factors such as total tannins, trypsin inhibitors, nitrate, oxalic acid and no cyanogenic glycosides, which favors consumption without absorptive maladjustments (MATHUR, 2005; SILVA et al. 2008; TEIXEIRA, 2012).

As for the vitamin composition of the leaves, Mathur (2005) states that they contain vitamin A, vitamin B1, vitamin B2, vitamin B3, vitamin C, calcium, chromium, copper, iron, magnesium, manganese, phosphorus, potassium and zinc. These micronutrients are essential for maintaining homeostasis in the body, as they have various beneficial properties that favor a state of balance. It could therefore be used as an adjunct to intervene in various nutritional deficiencies experienced in the country, such as iron and vitamin A deficiency, which in turn are issues related to public health. Powdered supplementation of the leaves would be one of the most effective ways of consuming it, always adding it to some preparation.

Regarding medicinal use, Razis, Ibrahim and Kntayya (2014) report that moringa has anti-fibrotic, anti-inflammatory, anti-microbial, anti-hyperglycemic, antioxidant, anti-tumor and anti-cancer properties. These activities make it a promising plant in this field. In this way, the consumption of moringa can be essential for the prevention of diseases in the population.

Therefore, the questions that gave rise to the study in question are associated with the search for the prevention of pathologies and improvement of the nutritional profile through low-cost supplementation; in addition to a new offer of a product of plant origin aimed at vegetarians; a possible intervention measure in cases of malnutrition and to contribute to further studies of this plant.

Given the wide range of nutrients available and the need for new research into the bromatological analysis of moringa leaves in different regions and climatic conditions, there was an interest in investigating the nutritional potential of its bioenergized leaves produced by a company in Paraná. On the other hand, there is a need for studies that explain the bioavailability of nutrients, especially proteins, *in vivo,* since the methods applied only look at total nitrogen content (TEIXEIRA et al., 2014; OLSON et al., 2016). This demonstrates the broad scope of new research into this problem.

In view of the above, this study has made a major contribution to the science of nutrition due to its investigation of the nutritional composition of bioenergized leaves, making it essential because it deals with a species of moringa grown differently from the others. As a result, it could lead to a new alternative form of food supplementation for the population, resulting in an improved quality of life based on healthy eating habits and physical exercise.

2 OBJECTIVES

2.1 General

To verify the nutritional potential of the bioenergized leaves of *Moringa oleifera* Lam. through bromatological analysis.

2.2 Specifics

Evaluate the pH, moisture, ash, protein, lipids, carbohydrates and energy value of the powdered leaves.

Compare the bromatological characteristics of the moringa under study with those of the same species described in the literature.

3 THEORETICAL FRAMEWORK

3.1 MEDICINAL PLANTS

Plants are a valuable object of study for researchers and scientists all over the world. This may be due to years of research into plants, in which many discoveries have been made about the therapeutic and nutritional properties that can or are applicable to human use. However, there are still countless species that need to be investigated, considering the diversity of plant species present in nature (BRASIL, 2006).

Traditional medicine is defined as the use of herbal products, as well as animal parts and minerals, the former being the most commonly used. Its aim is to preserve health or even treat physical or mental illnesses. It is characterized by a set of knowledge and practices taking into account theories, beliefs and experiences acquired from different cultures, although there are uncertainties about its effectiveness (WHO, 2013). It is believed that the use of plants for nutritional and therapeutic purposes has been applied since the beginning of humanity, mainly by the Egyptians, Greeks and Romans. Their premise was to promote health and treat illnesses in various populations. Great ancient scholars such as Hippocrates, Theophrastus, Celsus and Dioscorides already discussed the use of medicinal herbs (AZMIR et al., 2013).

In Brazil, the use of this practice is ancient and is also applied to health care, and its knowledge is transmitted orally between generations (FIGUEREDO; GURGEL; JUNIOR, 2014). According to the Ministry of Health (2006), Brazil is considered to be the richest nation in biodiversity in the world, with around 15 to 20% of this characteristic compared to other countries, as well as having different populations who know and apply traditional medicine. In view of its biodiversity, studies into the properties of plants occupy the first position in the ranking of research focusing on biomolecules, followed by fungi in particular as the second most studied biological material by researchers (BERLINCK et al., 2017).

Technological advances in the production of industrialized medicines led to a reduction in the use of herbs, which had been growing until the 20th century. The emergence of isolated substances and new syntheses of compounds from modern chemistry led to the use of medicinal plants being replaced by drugs synthesized in laboratories during the middle of the 20th century (FIGUEREDO; GURGEL; JUNIOR, 2014). On the other hand, in 1991, the WHO highlighted the contributions of traditional medicine, especially for individuals with little access to health systems, as well as encouraging the use of drugs extracted from natural products, with a view to their lower production costs compared to other medicines. Likewise, the organization stated that natural products, especially from plants, are promising for future therapeutic applications (BRASIL, 2006).

There is now recognition based on science that some medicinal plants, herbal medicines and the empirical knowledge of the population can be effective, as long as they are properly administered under professional prescription. On the other hand, the use of synthetic medicines to treat diseases is becoming less frequent due to their high costs, adverse effects and uncertain treatment responses (FIGUEREDO; GURGEL; JUNIOR, 2014).

Plants have compounds called bioactives, characterized as secondary metabolites that increase their ability to survive by generating adaptations to the environment. Bioactive compounds are also capable of promoting pharmacological or toxicological effects in humans and animals (AZMIR et al., 2013). These molecules are widely applied in research into organic synthesis and medicinal chemistry (BERLINCK et al., 2017).

Bioactive compounds can be found in various parts of the *Moringa oleifera* Lam species. They include ascorbic acid, β-carotene, quercetin, kaempferol and phenolic acids. The leaves in particular contain isothiocyanates, polyphenols and rutin. Its extracts, which are extracted using methanol and acetone, have antioxidant capacity (GUPTA et al., 2017). It is possible that the leaves of this tree could be a new input for the pharmaceutical and food industries in Brazil, as they combine both applications.

According to the Resolution of the Collegiate Board (RDC) No. 26, of May 13, 2014, of the National Health Surveillance Agency (ANVISA), moringa is classified as a traditional herbal product, which is a product made from active plant raw materials in which the literature assures efficacy and safety, as well as not requiring a prescription and medical monitoring for use. In addition, RDC No. 27, of August 6, 2010, characterizes moringa as a product exempt from health registration, according to code 4200098, which contains the exemption for "mixtures for the preparation of food and ready-to-eat food".

3.2 BIOENERGIZING FOOD

The current agricultural system is deficient in aspects of plant metabolism. Nitrogen, phosphorus and potassium are the three nutrients that promote plant development. However, it has been observed that these are insufficient for their metabolism, making them susceptible to disease and pest attack, which in turn leads to the use of agrochemicals. To ensure optimum growth, plants and pastures need more than 30 micronutrients to become nutritious food for consumption. However, the nutrients that are essential for plant growth have been transferred from the soil to the sea due to the occurrence of plantations and erosion. In addition, another factor that adds to the lack of nutrients in vegetables is the low composition of trace elements in commercial fertilizers, limited to the addition of a maximum of 6 of these elements (FILHO, 2017).

According to Filho (2017), the practices of natural agriculture are not sufficient to meet the nutritional needs of plants, although they are superior to those of conventional agriculture. In addition, adding mineral salts directly to the food would have no absorptive effect, since for biological reasons, the consumption of nutrients only from vegetables leads to the conversion of inorganic elements into organic ones in the body. On the other hand, studies show that the availability of nutrients is more important than the amount added to the plants. In this sense, the balance of minerals contained in seawater provides the nutrients that plants need for their metabolism and bioavailability to humans. Therefore, bioenergizing consists of adding these minerals to food in order to enhance their assimilation and meet the body's nutritional needs.

3.3 *Moringa oleifera* Lam.

3.3.1 BOTANICAL CLASSIFICATION

Moringa oleifera Lamarck or *Moringa pterygosperma* Gaertner is a leguminous plant in the Moringaceae family, popularly known in Brazil as moringa, white lily or okra. Moringaceae is a monogeneric family containing 13 species. they are: *M. arborea, M. rivae, M. borziana, M. pygmaea, M. longituba, M. stenopetala, M. ruspoliana, M. ovalifolia, M. drouhardii, M. hildebrandi, M. peregrine, M. concanensis and Moringa oleifera* (FERREiRA, et al. 2008; TEIXEIRA, 2012; LEONE, et al. 2015; BRILHANTE, et al. 2017). Table 1 shows the taxonomic classification of the moringa studied.

TABLE 1: Taxonomic classification of *Moringa oleifera* Lam.

SCIENTIFIC CLASSIFICATION
Kingdom: Plantae
Division: Magnoliophyta
Class: Magnoliopsida Order: Brassicales Family: Moringaceae Genus: Moringa Species: *Moringa oleifera*

Binomial nomenclature: *Moringa oleifera* Lam.

SOURCE: Adapted from KARIM et al. (2016).

The tree comes from northern India but is widely found in America, Africa, Europe, Oceania and Asia (Figure 1). It is characterized by deciduousness and rapid growth regardless of soil conditions, as well as a wide pH range (5.0 - 9.0), with a favorable temperature of between 26 °C and 40 °C. It consists of leaves (Figure 2), flowers (Figure 3), pods (Figure 4), roots (Figure 5) and seeds (Figure 6), producing fruit and seeds all year round. The tree can reach heights of between 5 and 15m and has a thorax diameter of between 20cm and 40cm, with light-colored, spongy bark. Its leaves are feathery and pale green in color, ranging in length from 30cm to 60cm. The leaves are white or cream in color and have a characteristic scent, with a diameter of 2.5cm. The fruit splits into three parts when dry. The seeds are dark brown and are divided into three paper-like structures. The main root is very thick (RANGEL, 1999; ANWAR, et al., 2007; FERREIRA, et al. 2008; BRILHANTE, et al. 2017; GUPTA, et al. 2017).

FIGURE 1: *Moringa oleifera* Lam

SOURCE: The author

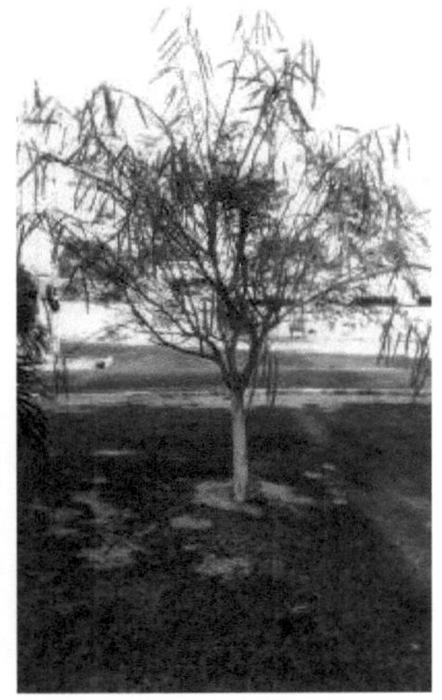

FIGURE 2: Leaves

SOURCE: The author

FIGURE 3: Flowers

SOURCE: The author

FIGURE 4: Pods

SOURCE: The author

FIGURE 5: Roots

SOURCE: The author

FIGURE 6: Seeds

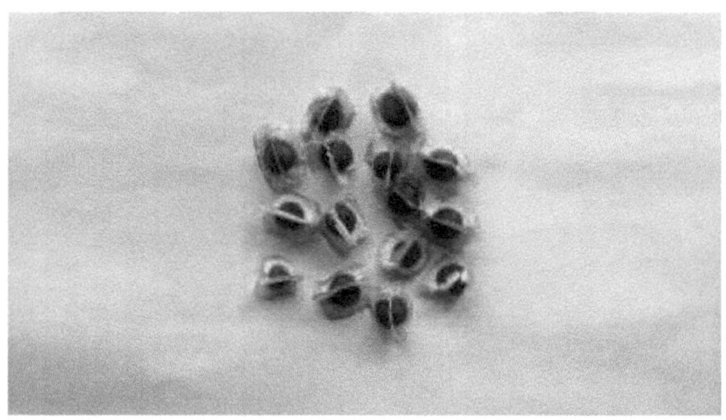

3.3.2 NUTRITIONAL COMPOSITION

M. oleifera contains essential nutrients such as vitamins, minerals, amino acids, antioxidant and anti-inflammatory compounds, as well as omega 3 and 6 fatty acids. It is suggested that the leaves may contain vitamin A through β-carontene, B-complex vitamins, vitamin C, D and E. The minerals present are: calcium, potassium, zinc, magnesium, iron and copper. It also contains tannins, sterols, terpenoids, flavonoids, saponins, anthraquinones and alkaloids as phytochemical agents. Anti-cancer agents such as glucosinolates, isothiocyanates, glycosidic compounds and glycerol-1-9-octadecanoate also stand out (RAZIS; IBRAHIM; KNTAYYA, 2014; GOPALAKRISHNAN, DORIYA, KUMAR, 2016).

The calorie, macronutrient and fiber values (Table 2), aminogram (Table 3) and vitamin and mineral content (Table 4) of moringa leaves were demonstrated by Gopalan (1971) and Fuglie (2002).

TABLE 2: Values per 100 grams of nutrients in the leaves of *Moringa oleifera* Lam.

NUTRIENTS	FRESH LEAVES (GOPALAN, 1971)	DRIED LEAVES (FUGLIE, 2002)
Calories	92 Kcal	205 Kcal
Carbohydrates	12,5 g	38,2 g
Proteins	6,70 g	27,1 g
Lipids	1,70 g	2,3 g

Fibers	0,90 g	19,2 g

SOURCE: Adapted from Gopalan (1971); Fuglie (2002).

TABLE 3: Values per 100 grams of amino acids in the leaves of *Moringa oleifera* Lam.

AMINO ACIDS	FRESH LEAVES (GOPALAN, 1971)	DRY LEAVES (FUGLIE, 2002)
Arginine	406.6 mg	1,325 mg
Histidine	149.8 mg	613 mg
Isoleucine	299.6 mg	825 mg
Leucine	492.2 mg	1,950 mg
Lysine	342.4 mg	1,325 mg
Methionine	117.7 mg	350 mg
Phenylalanine	310.3 mg	1,388 mg
Threonine	117.7 mg	1,188 mg
Tryptophan	107 mg	425 mg
Valine	374.5 mg	1.063 mg

SOURCE: Gopalan (1971); Fuglie (2002).

TABLE 4: Values per 100 grams of portion of vitamins and minerals in the leaves of *Moringa oleifera* Lam.

VITAMINS AND MINERALS	FRESH LEAVES (GOPALAN, 1971)	DRY LEAVES (FUGLIE, 2002)
Carotene (Vit. A)	6.78 mg	18.9 mg
Thiamine (B1)	0.06 mg	2.64 mg
Riboflavin (B2)	0.05 mg	20.5 mg
Niacin (B3)	0.8 mg	8.2 mg
Vitamin C	220 mg	17.3 mg
Càlcio	440 mg	2.003 mg

Copper	0.07 mg	0.57 mg
Iron	0.85 mg	282 mg
Magnesium	42 mg	368 mg
Phosphorus	70 mg	204 mg
Potassium	259 mg	1.324 mg
Zinc	0.16 mg	3.29 mg

SOURCE: Adapted from Gopalan (1971); Fuglie (2002).

Its bioactive compounds are responsible for numerous medicinal applications. Vitamins, phenolic acids, flavonoids, isothiocyanates, tannins and saponins play a fundamental role in this function. Studies show that these are responsible for promoting hypolipidemic, antioxidant, anti-inflammatory, immunomodulatory, hepatoprotective, antihyperglycemic, hypotensive, anticancer and anti-neurodegenerative effects. Therefore, the consumption of moringa leaves is a strategy that can guarantee prevention and protection against various diseases, as well as boosting the consumption of essential nutrients in the body (VERGARA-JIMENEZ; ALMATRAFI; FERNANDEZ, 2017).

3.3.3 ANTIOXIDANT POTENTIAL

Living beings are capable of processing numerous chemical compounds for their maintenance and survival. The compounds in biological systems are divided into primary and secondary classes. The primary class includes the metabolites involved in growth and development, such as carbohydrates, amino acids, proteins and lipids. The secondary class contains compounds that play a protective role for plants, with unusual chemical structures that can be separated, identified and characterized using appropriate extraction techniques. They are divided into three main classes: terpenes and terpenoids, alkaloids and phenolic compounds (AZMIR et al., 2013).

In the case of these compounds, it is possible to identify secondary metabolites with potential therapeutic effects in *M. oleifera*. These include alkaloids, tannins, flavonoids, steroids, saponins, coumarins, quinones and resins. The leaves of the plant contain kaempferol in particular, a flavonoid that promotes nutritional, anti-inflammatory and antimicrobial activities and can be extracted via ethanolic extract. Flavonoids in turn act as antioxidants by protecting the main biomolecules against oxidative stress, a process which is responsible for the appearance of various pathologies (BRILHANTE et al., 2017; MARTiNEZ-GONZALEZ et al., 2017).

Studies have shown the antioxidant capacity of the aqueous extract of the leaves of this plant in rats subjected to radiation-induced toxicity, using the DPPH (2,2-diphenyl-1-picrylhydrazyl) radical method. This benefit is justified by the high content of antioxidants found in this extract, which provide protection against oxidative damage. It may have pharmacological potential, acting to prevent cardiovascular and pulmonary diseases (MANSOUR; ISMAEL; HAFEZ, 2014). The antioxidant properties of the leaves of this plant are also evaluated using the ABTS (2,2-azinobis-3-ethyl-benzothiazoline-6-sulfonic acid) method. This method has identified the presence of polyphenolic compounds that protect against diseases induced by oxidative stress (MOYO et al., 2012).

3.3.4 ANTI-MICROBIAL ACTIVITY

With regard to bacteria, it is known that they are closely related to the human body, since the *Bacillus cereus* genus causes food poisoning, while *Escherichia coli, Staphylococcus aureus* and *Pseudomonas aeruginosa* cause abortions and respiratory complications. With a view to attributing *M. oleifera* as a possible pharmacological resource, research shows that the aqueous extract of the leaves has an antibacterial effect on Gram-positive bacteria of the species *B. cereus, E. faecalis, S. aureus and S. epidermidis*; and Gram-negative bacteria of the classes *S. enterica, S. typhi, P. aeruginosa, P. vulgaris, E. cloacae, E. coli* and *K. pneumoniae. Pneumoniae*. Although it shows pharmacological potential against microbial pathogens, more studies are needed on the isolation of secondary metabolites from the extract to explain its action on bacterial metabolic pathways (AL_HUSNAN; ALKAHTANI, 2016; BRILHANTE et al., 2017).

Another class of microorganisms that can have adverse effects on the body are fungi. They are ubiquitous beings found anywhere in the world, which when inhaled cause respiratory complications such as allergic bronchopulmonary mycosis, allergic fungal sinusitis and hypersensitivity pneumonitis. With regard to medicinal use, the aqueous extract of *Moringa has* demonstrated antifungal action on the species *Aspergillus niger; A. flavus, Alternaria spp., Fusarium spp. R. stolonifer, Penicillium sp., Candida albicans, C. glabrata, C. tropicalis, Trichophyton rubrum, T. mentagrophytes, Epidermophyton floccosum* and *Microsporum canis*. Its antimicrobial activity comes from the benzyl isothiocyanate and flavonoids present in the seed and leaf extracts (BAXI et al., 2016; AL_HUSNAN; ALKAHTANI, 2016; BRILHANTE et al., 2017).

3.3.5 ANTI-LEISHMANIAL CAPACITY

From a Brazilian epidemiological perspective, the WHO considers leishmaniasis to be the most neglected disease. The term "neglected" is based on the lack of research into the production of new drugs, vaccines, among other technologies that favor the prevention and treatment of leishmaniasis.

Another view that corroborates the term is based on the insufficient attention given to the disease in the health services and in the political sphere. The disease caused approximately six thousand deaths in Brazil in 2008, along with malaria, Chagas disease and schistosomiasis (WERNECK; HASSELMANN; GOUVÊA, 2011).

Leishmaniasis manifests itself in cutaneous, muco-cutaneous and visceral forms, covering various signs and symptoms in the body. The drugs used to treat it are expensive and have considerable side effects, making it difficult for the poorest populations to access them. With this in mind, research into the use of natural products as a treatment has been investigated. With regard to the use of these products, *M. oleifera has been* suggested as an adjuvant for the treatment of leishmaniasis, since numerous compounds with therapeutic characteristics can be found in the tree. Never studied for this purpose, it was recently discovered that some extracts of the roots and leaves evaluated *in vitro* showed antileishmanial activity against the *Leishmania donovani* species. In the future, the use of the plant could lead to a new therapy for the treatment of the disease, given the alarming WHO records that reveal 12 million people infected and 350 million at risk of infection (KAUR et al., 2014).

3.3.6 DOSAGE AND TOXICITY

Regarding the safe dosage for the use of moringa leaves, it is known to date that research is only directed at rats, considering the ingestion of aqueous and methanolic extracts of the leaves. With this in mind, Awodele et al. (2012) carried out an acute toxicity study on Wistar rats. 250, 500 and 1500 mg/kg of the aqueous extract were administered orally for 60 days and the rats in the control group were given distilled water. To observe the level of toxicity, sperm quality, hematological and biochemical aspects such as liver enzymes, urea and creatinine were analyzed, as well as a histopathological examination. At the end of the study, the authors concluded that there were no significant changes in sperm quality, hematological and biochemical parameters in the group given the extract compared to the control group.

Asare et al. (2012) also investigated the toxicity of moringa in the form of an aqueous extract in rats. The study was divided into 3 groups, in which the first ingested low doses (1000mg/kg), the second high doses (3000mg/kg), and the third corresponded to the control group, which consumed a saline solution. The biochemical and hematological aspects were analyzed over 14 days. The extract proved to be toxic in the group that received the high doses, while the low doses were not harmful, making them safe.

From another point of view, moringa showed toxicity in rats that consumed the methanolic extract of the leaves, which could predispose to liver and kidney damage when consumed chronically (OYAGBEMI et al., 2013). Although there is disagreement about the safety of acute use compared to chronic use, it is understood that the dosages for administration in humans have not been fully elucidated, given that the literature covers a small number of studies aimed at them.

Some short-term interventional studies in humans describe the use of daily doses of between 4.6 g and 8 g of the powdered leaves. The plant has had a positive impact on diabetes, dyslipidemia and antioxidant capacity without promoting adverse effects (STOHS; HARTMAN, 2015). Although there is no consensus on dosage and toxicity, Adedapo, Mogbojuri and Emikpe (2009) report that moringa leaves are relatively safe for nutritional and pharmacological use.

4 METHODOLOGY

This study explains the methods applied to the bromatological study in order to make it easier for the reader to understand the procedures carried out. For the theoretical basis, articles and books from online databases were used: Google Scholar®, PubMed Central®, Scielo® (Scientific Electronic Library Online) and ScienceDirect®. The descriptors used in the search were: "moringa oleifera lam", "bromatological analysis", "moringa and nutrition", "moringa and health", "physical-chemical analysis", "moringa and nutrition", "moringa and health".

The inclusion criteria were original articles, reviews and course final papers related to the application of moringa in malnutrition, the chemical analysis of its leaves and its use for nutritional and pharmacological purposes. Works related to moringa in animal feed (except on toxicity) and on the use of other parts of the plant were excluded.

4.1 METHODOLOGICAL APPROACH

The research in question, according to the Coordination for the Improvement of Higher Education Personnel (CAPES), falls within the scope of Health Sciences, with biochemistry of nutrition as its sub-area. The approach to the work is quantitative, qualitative, exploratory, descriptive, experimental, documental, pure and bibliographical. These methods add a wide range of information to the study, resulting in data that leads to concrete conclusions.

Quantitative research has principles aimed at finding data, presenting conclusions based on numbers. Qualitative research, on the other hand, investigates the essence of the content by looking at its variables. In short, quantitative studies use data grouped in tables, while qualitative studies use texts, matrices, diagrams, among other means. From another perspective, pure research aims to contribute to the evolution of science regardless of its applications (GIL, 2002; GIL, 2008).

As it was carried out entirely in a laboratory, this study falls within the scope of experimental research, as it fully evaluates the sample. Gil (2002) argues that experiments offer the possibility of control, ensuring that the research is more reliable than other designs. In this sense, the aim of this study is to analyze *Moringa oleifera* leaves globally by means of bromatological analysis.

4.2 METHODOLOGICAL PROCEDURES

The tests for the bromatological analysis of the bioenergized moringa leaves took place at the Biochemistry and Bromatology Laboratory of UNINASSAU, located in the city of Parnaiba, state of Piaui. The leaves of the plant were grown and encapsulated in the same city and provided by professor and researcher Francisco Edson Teófilo Filho, director of the company VitalBio[®] - Alimentos Naturais Bioenergizados.

4.3 BROMATOLOGICAL STUDY

The methodology of the Adolf Lutz Institute (IAL, 2008) was used because it is

This method has been validated and widely used in other analytical studies in the literature. Based on this, the pH, moisture, ash, protein and lipid values of the sample were determined. The carbohydrate and calorie contents were determined according to the methodologies of Gonçalves (2006) and Osborne and Voogt (1978) respectively. All procedures were carried out in triplicate to ensure a smaller margin of error in the data, and the mean and standard deviation were then calculated. To carry out the tests, it was necessary to remove the leaf powder contained in the capsules (Figure 7).

FIGURE 7: Sample removed from capsules

SOURCE: The author

4.4 pH DETERMINATION

The purpose of pH determination is to investigate the activity of hydrogen ions through a potentiometric measurement using a glass electrode and a reference electrode, or combined electrode. The glass electrode system measures the electromotive force varying linearly with pH. The pH measuring equipment consists of the combined electrode, using buffer solutions of known pH to calibrate it (IAL, 2008).

- Materials and equipment

-       *Moringa oleifera* Lam. leaves in powder form;

-       Filter paper;

-       pH meter - Tecnal[®] Microprocessed Tec-11;

-       Magnetic stirrer without heating - Quimis® - Q221mag;

-       Magnetic bar;

-       50 mL and 150 mL beakers.

- Reagents and solutions

-       Biftalate buffer solution pH 4 at 25 °C - Tecnal®;

-       Phosphate buffer solution pH 7 at 25 °C - Tecnal® ·

- Procedure

The instrument was first calibrated according to the recommendations in the manual. The glass electrode was washed with distilled water, then carefully dried with fine absorbent paper. The electrode was then inserted into the phosphate buffer solution prepared in the 50 mL beaker, stirring gently and constantly. The temperature of the solution was checked together with the pH value of the buffer. The electrode was then removed from the solution and washed with distilled water. The bi-phthalate cap was then used to adjust the apparatus, which is used for samples with a pH < 7.

To determine the pH, approximately 5 g of the sample was weighed into a 150 mL beaker. Then 100 mL of distilled water was poured into the beaker along with a magnetic bar for dilution. It was then placed on the magnetic stirrer. The device was set to 1290 rpm for the three samples for 15 minutes. After the procedure was finished, the beakers were removed after 5 minutes. At the end, the pH of the samples was checked by first washing the electrodes with distilled water and then gently inserting them into the sample beaker. The pH was read and recorded after the value remained constant.

4.5 HUMIDITY ASSESSMENT

Moisture corresponds to the fraction of water contained in food. According to Gonçalves (2006), moisture data is used to classify foods as perishable, semi-perishable and non-perishable, in addition to promoting chemical and biochemical events in foods. It can be said that humidity is a predictor of food durability, and it is essential to analyze it in order to characterize its perishability.

- Materials and equipment

- *Moringa oleifera* Lam. leaves in powder form;

- Electronic analytical balance - Bioprecisa® JA3003N;

- Desiccator with silica gel;

- Greenhouse - Fanem® 515;

- Porcelain capsule.

- Procedure

Three empty crucibles were weighed using an analytical balance. The values were recorded immediately. Then, previously weighing the weight of each crucible, 2 g of the sample was inserted. They were then placed in the oven where they were heated for 1 hour at 130 °C. At the end of the heating time, the crucibles were removed to be cooled in a desiccator so as not to cause weight variations. The crucibles were brought to room temperature after 30 minutes, and then placed back on the analytical balance to be weighed again. The constant weight values of each crucible were recorded.

- Calculation

$\underline{100 \times N}$ = humidity at 130 °C percent m/m

P

Where:

N = number of grams of moisture

P = number of grams of sample

4.6 ASH ANALYSIS

The residue obtained by heating a product at temperatures close to 550 °C to 570 °C is called incineration residue or ash, corresponding to the portion that consists of the inorganic matter of the food. It can be extracted using high temperatures or by applying strong acids or bases, which in turn destroy the organic matter (GONÇALVES, 2006; IAL, 2008).

- Materials and equipment

- *Moringa oleifera* Lam. leaves in powder form;

- Crucible;

- Microprocessor muffle furnace - Quimis®Q318S;

- Desiccator with silica gel;

- Pyroceramic platform heating plate - Cienlab® CE-200/P1;

- Electronic analytical balance - Bioprecisa® JA3003N;

- Crucible tweezers.

• Procedure

The remaining samples from the moisture analysis were used in this test. The crucibles were then placed on the heating plate for 15 minutes. They were then placed in the muffle furnace, which had previously been heated to 550 °C, and left for 2 hours for the incineration process. After the waiting time, the crucibles were transferred from the muffle furnace to the desiccator using tweezers. The samples appeared slightly grayish, as recommended in the literature. They were removed after 30 minutes. They were then weighed on an analytical balance and the values recorded.

• Calculation

$\underline{100 \times N}$ = percent ash m/m

P

Where:

N = number of grams of ash P = number of grams of sample

4.7 PROTEIN DETERMINATION

Proteins are molecular structures made up of amino acids. They are responsible for the nutritional quality of food by providing essential amino acids (GONÇALVES, 2006). In living systems, they are the most abundant class of molecules and play a fundamental role in vital processes (HARVEY; FERRIER, 2012).

- Materials and equipment

- *Moringa oleifera* Lam. leaves in powder form;

- Filter paper;

- Electronic analytical balance - Bioprecisa® JA3003N;

- Test tube;

- Microprocessed Kjeldahl Microdigester Block - Quimis® Q327M;

- 300 mL Kjeldahl flasks;

- Heating blanket - Quimis® ;

- 250 mL Erlenmeyer flask;

- 25 mL burette.

- Reagents and solutions

- Sulfuric acid P.A;

- Catalytic mixture: potassium sulphate (K_2SO_4) and copper sulphate pentahydrate ($CuSO_4.5H_2O$) P.A - 10:1;

- 50% sodium hydroxide (NaOH);

- Boric acid (H_3BO_3) at 4%;

- Mixed indicator: 0.132 g methyl red ($C_{15}H_{15}N_3O_2$), and 0.06 g bromocresol green ($C_{21}H_{14}Br_4O_5S$);

- Hydrochloric acid (HCl) 0.1N.

• Procedure

The first stage of protein determination consisted of the digestion process. To begin this procedure, 2 g of the sample and 2.5 g of the catalyst mixture were weighed on filter paper on an analytical balance. These elements were inserted into the test tubes. After placing the tubes in the microdigester installed inside the chapel, 20 mL of sulfuric acid P.A. was added to each tube. The equipment was turned on to promote gradual heating until it reached 400 °C, totaling 3 hours of analysis. At the end of the process, the already-digested samples were in liquid form, with a clean, transparent appearance and a blue-green color. The tubes were carefully removed to cool naturally.

The second stage involved distilling the samples. 20 mL of 4% boric acid with 4 to 5 drops of mixed indicator solution was poured into the erlenmeyer flasks. The liquids from the digestion process were transferred to the Kjeldahl flasks, followed by 10 mL of distilled water. A 50% sodium hydroxide solution was then prepared and about 25 mL was added to the tubes where it reached a dark color. The tubes were fitted to the heating mantle and distiller and heated to boiling. The receiving solution was kept cold during distillation.

Titration was the last stage of the test. 25 mL of 0.1N hydrochloric acid was poured into the burette and the receiving solution was titrated until the indicator turned. The volume of titration used was then recorded.

- Calculation

$$\% \text{ total nitrogen} = \frac{V \times N \times f \times 0.014 \times 100}{P}$$

% protein = % total nitrogen x F

Where:

V: milliliters of 0.1 N sulfuric acid solution or 0.1 N hydrochloric acid solution used in the titration, after correcting the blank;

N: theoretical normality of the 0.1N sulfuric acid solution or 0.1N hydrochloric acid solution;

f correction factor of 0.1N sulfuric acid solution or 0.1N hydrochloric acid solution;

P: sample mass in grams;

F: nitrogen/protein ratio conversion factor, according to the product.

Meat and meat products = 6.25

Gelatine = 5.55

Milk and dairy products = 6.38

Wheat and wheat products = 5.7

Eggs = 6.68

Rice =5,95

Soy = 5.71

Barley, oats, rye = 5.83

Walnuts = 5.46

4.8 lipid analysis

Lipids are high-energy organic compounds that contain essential fatty acids for the body and act as transporters of fat-soluble vitamins. They are usually extracted from food using organic solvents such as ether, using the Soxhlet apparatus (IAL, 2008).

- Materials and equipment

-        *Moringa oleifera* Lam. leaves in powder form;

-        Crucible;

-        Filter paper;

-        Greenhouse - F anem® 515;

-        Soxhlet extractor with 300 mL flat-bottomed flask;

-        Soxhlet extractor - Quimis® Q308G;

-        Electronic analytical balance - Bioprecisa® JA3003N;

-        Heating plate - Fisatom® 503-1;

-        Desiccator with silica gel.

- Reagents and solutions

- Petroleum ether P.A.

•   Procedure

Using a pre-weighed analytical balance, 2 g of the sample was weighed into the crucible, as well as the empty Soxhlet extractor container without the fitting rim. The sample was then placed in an oven at 130 °C for 40 minutes to remove any residual moisture. After heating, the sample was transferred to the filter paper. The filter paper was then placed in the extractor and enough ether was added for one and a half Soxhlets, keeping it under continuous heating on the hotplate.

After about 2 hours of the procedure, the flat-bottomed flask was removed and the solvent in it was transferred to the container of the Soxhlet extractor. At this stage, the solvent was only heated in the apparatus until it evaporated. The container was then removed and placed in an oven at 105 °C for 1 hour. It was then removed and placed in the desiccator to cool for 30 minutes. At the end, it was placed on the analytical balance to be reweighed.

•   Calculation

$\underline{100 \times N}$ = lipids or ethereal extract percent m/m

P

Where:

N = no. of g of lipids   P = no. of g of sample

4.9 TOTAL CARBOHYDRATES

The most bioabundant biomolecules on earth are carbohydrates. Sugar and starch are the main elements that make up diets in different parts of the world (NELSON; COX, 2014). The glycemic fraction of the food is determined by subtracting the sum of the nutrients that represent 100% of it. This result can include dietary fiber, which when analyzed separately is added to the total nutrient values (GONÇALVES, 2006).

- Calculation

TOTAL GLYCDIES = 100 - (% moisture + % ash + % protein + % lipids)

4.10 ENERGY VALUE

The energy value of the sample was determined using the nutrient conversion factors according to the methodology of Osborne and Voogt (1978). Carbohydrates and proteins were given 4 kcal/g and lipids 9 kcal/g. The total value was presented in kcal/100g.

- Calculation

VE = (% proteins x 4) + (% lipids x 9) + (% carbohydrates x 4)

Where:

VE: Energy value

5 Results and discussion

5.1 BROMATOLOGICAL TEST

The results of the bromatology test on moringa leaves, along with the mean, standard deviation and calorific value, are shown in Table 5, according to the sequence described in the methodology. The values of the bioenergized moringa were compared with the data found in the literature.

TABLE 5: Results of the bromatological analysis of bioenergized moringa leaves.

ANALYSIS	A. 01	A. 02	A. 03	AVERAGE	D. P.	V. E. (Kcal)
pH	5,250	5,280	5,290	5,270	0,020	-
Humidity	31,450	34,850	35,150	33,810	2,055	-
Ash (g/100g)	8,600	8,500	8,600	8,566	0,057	-
Protein (g/100g)	7,040	10,911	9,134	9,028	1,937	36,112
Lipids (g/100g)	9,714	7,407	7,502	8,207	1,305	73,863
Carbohydrates (g/100g)	43,196	38,332	39,614	40,380	2,521	161,520
TOTAL						271,506

LEGEND: A.: Sample; D. P.: Standard deviation; V. E.: Energy value (caloric).

The pH measurement corresponds to the concentration of hydrogen ions in the food. It is essential because it has an impact on the conditions in which microorganisms develop (GAVA, 1984). The samples were slightly acidic, with values close to those found by Teixeira (2012). In her study, the author evaluated the chemical composition of the flour obtained from *Moringa oleifera* Lam. leaves, obtaining pH values between 5.48 and 5.90, using the methodology based on adding electrodes to the sample, which is similar to the procedure used in this work. It can therefore be understood that the results are in line with the literature, although there could be possible variations in view of the different regions and cultivations of the plant. Graph 1 shows comparisons between the sample studied and other studies.

CHART 1: pH range of the sample analyzed compared to the study

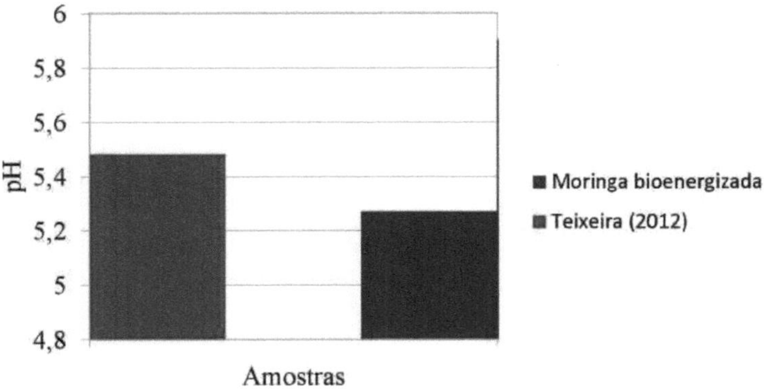

Samples from Teixeira (2012).

Moisture levels are closely associated with the shelf life of the product, taking into account nutritional properties (CELESTINO, 2010), physical modifications and the multiplication of microorganisms (FRANCO; LANDGRAF, 2004). Graph 2 shows the greatly increased levels of humidity compared to the studies by Gasqui et al. (2013), in which 5.7% was obtained, and Teixeira (2012), which amounted to 9%.

GRAPH 2: Moisture content of the moringa investigated in relation to other authors.

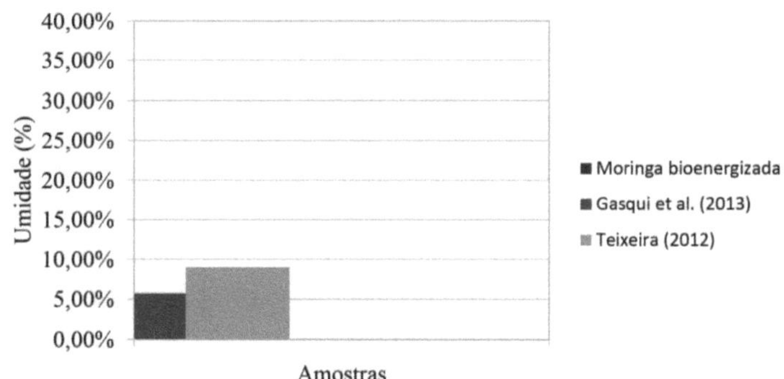

Research involving moringa does not indicate standard values for moisture content, and there have been limited studies on this analysis. According to Franco and Landgraf (2004), foods with a moisture content of between 15% and 50% are considered *intermediate* moisture foods (IMF). In this class of products, humectant and fungistatic additives are used to improve the quality of the food. In this sense, the sample studied can be included in this classification according to the results found.

The ash found, although low, was close to the values obtained in the analysis by Leone et al.

(2015), in which they investigated samples from three different countries: Chad (10.79 g), Algeria (13.38 g) and Haiti (9.62 g), according to Graph 3. Based on the data, it is suggested that there is a good proportion of micronutrients in the sample investigated. However, future research involving the quantification of these elements separately is needed.

GRAPH 3: Number of ashes in the moringa under study compared to samples from three different countries.

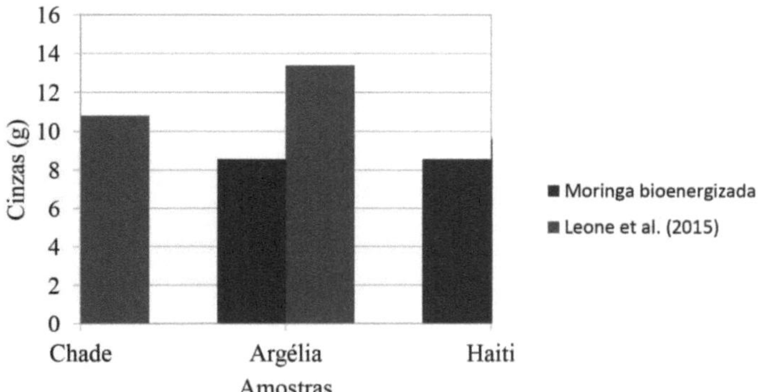

In terms of protein content, there is a significant contrast between the sample from the present study and those from other authors. In this respect, Moyo et al. (2011) obtained 30.29 g in their analysis, and Yaméogo et al. (2011) 11.9 g. The different proportions of protein and other nutrients between the samples is a possible reflection of the local adaptations undergone by the plant after its cultivation outside the Indian continent (MATHUR, 2005). Graph 4 shows the protein values.

GRAPH 4: Protein content of the moringa studied compared to other research.

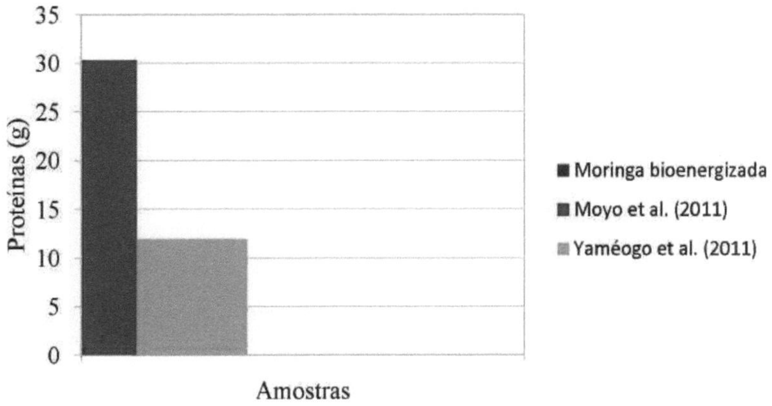

At the end of the first stage of protein analysis, one of the samples was not completely digested,

31

which meant that another one had to be prepared to maintain the standard of the study in triplicate. Interestingly, it was found that in acidic media the protein acquired a gel-like structure, occupying the lower part of the test tube. It is assumed that the leaves analyzed contain a considerable amount of dietary fiber, since they are not digested by enzymes (MAHAN; ESCOTT-STUMP; RAYMOND, 2013). Figure 8 shows the sample at the end of the digestion process.

FIGURE 8: Undigested sample

SOURCE: The author

The results for lipids are in line with the findings in the literature. The values were moderately higher than those of Teixeira et al. (2014), who obtained 7.09 g, and Shih et al. (2011), who obtained 5.75 g (Graph 5). As well as naturally having an excellent fatty acid profile, especially the essential ones (SAINI; SIVANESAN; KEUM, 2016), there is a good proportion of lipids in the leaves studied in this work.

GRAPH 5: Relationship between lipids in the moringa analyzed and studies by other authors.

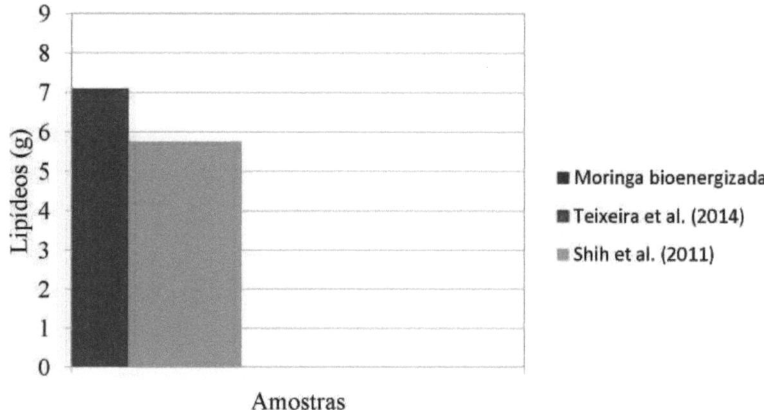

The data in the table shows that the carbohydrate content is the highest, which could be explained by the fact that the sample is of vegetable origin and may also be a good source of fiber, as shown in the work by Teixeira (2012), which found 33.37 g of carbohydrates and 10.99 g of fiber. Still on the subject of carbohydrates, Gasqui et al. (2013) found 38.8 g in their tests. It is therefore possible to see that the moringa under study has a relatively higher number than that obtained by the two authors mentioned, as shown in Graph 6. However, in order to make concrete statements about fiber, it would be necessary to carry out a further study analyzing its content in isolation.

GRAPH 6: Carbohydrate value of the sample compared to other studies.

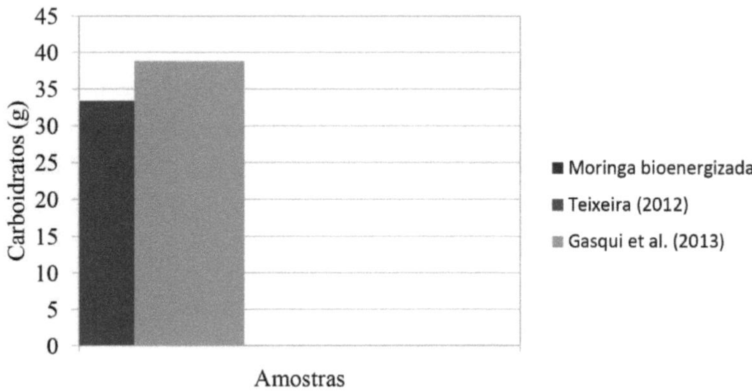

The calories identified in the sample were higher than those found by Gopalan (1971) and Fuglie (2002), 92 Kcal and 205 Kcal respectively (Graph 7). It is worth noting that the ratio between macronutrients affects the energy value of the food, as there are differences between nutrients in research. Considering a daily consumption of between 4.6 g and 8 g of leaf powder (STOHS;

HARTMAN, 2015), an average of 17.12 Kcal is obtained from the sample analyzed, representing a derisory amount of energy for the individual. Therefore, when taken with caution, it can help with weight loss strategies as well as maintaining health or preventing diseases.

 Number of calories found in this study and by other authors.

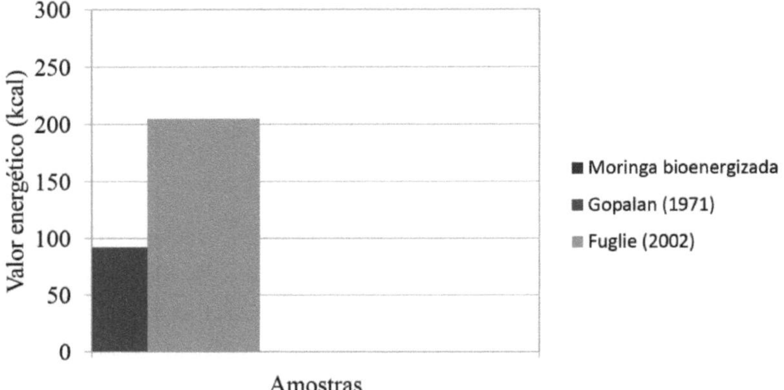

Based on the analysis of studies related to the bromatological characteristics of moringa leaves, it can be seen that overall there were few significant differences between the sample investigated and data from the literature. It is assumed that the large differences found in protein and moisture content are due to various factors, mainly environmental. Higher results were expected, especially in the protein profile, given that the plant is intended to be grown for bioenergizing purposes.

However, the leaf powder showed a good proportion of nutrients, consolidating it as an alternative food supplement for various purposes. Although the bromatological analysis shows some limitations in terms of nutrient specificity, it was possible to identify that the sample has nutritional potential because it is similar to the data recommended in the literature.

6 FINAL CONSIDERATIONS

Faced with countless pathologies to be treated, there is a lack of new plant-based products that can help prevent and treat them. Over the years, moringa has generated greater attention among scholars due to the numerous discoveries about its therapeutic and nutritional properties, which have consequently acquired relevance in modern science. These reasons led to the production of this study.

Through the tests, the nutritional potential of the leaves studied was verified in relation to the bibliographic data *already* recommended through the similarities between most of the analyses. Bioenergization therefore plays a fundamental role in improving the nutritional value of vegetables. However, further molecular research is essential to elucidate further nutritional and medicinal approaches to moringa cultivated in this way.

The plant's nutritional and pharmacological uses include the control of free radicals and the reduction of diseases resulting from this condition, as well as its use against bacteria and fungi, as a new alternative to the use of antimicrobials, and as a nutritional supplement for healthy individuals, sports enthusiasts, the infirm, and especially vegetarians, as it provides a good supply of essential nutrients for this public.

As a result, this work will provide new perspectives for accessible treatments that can help economically disadvantaged populations to combat nutritional deficiencies and neglected diseases, as well as adding more scientific information to the health sciences, especially in the field of nutrition science.

7 BIBLIOGRAPHICAL REFERENCES

ADEDAPO, A. A.; MOGBOJURI, O. M.; EMIKPE, B. O. Safety evaluations of the aqueous extract of the leaves of *Moringa oleifera* in rats. **Journal of Medicinal Plants Research**, Vol.3, p.586-591, 2009.

AL_HUSNAN, L. A.; ALKAHTANI, M. D. F. Impact of Moringa aqueous extract on pathogenic bacteria and fungi *in vitro*. **Annals of Agricultural Science**, v.61, p.247-250, 2016.

ANVISA. **Resolution RDC No. 26, of May 13, 2014**. Ministry of Health - MS.

National Health Surveillance Agency - Anvisa. Available at:

<http://portal.anvisa.gov.br/documents/33836/351410/Consolidado+de+normas+da+COFI
D+%28Vers%C3%A3o+V%29/3ec7b534-a90f-49da-9c53-ce32c5c6e60d>. Accessed on: 12 Nov. 2017.

ANVISA. **Resolution RDC No. 27, of August 6, 2010**. Ministry of Health - MS. National Health Surveillance Agency - Anvisa. Available at: <http://portal.anvisa.gov.br/en/registros-e-autorizacoes/alimentos/produtos/isencao-de- registro>. Accessed on: November 12, 2017.

ANWAR, F.; LATIF, S.; ASHRAF, M.; GILANI, A. H. *Moringa oleifera*: A Food Plant with Multiple Medicinal Uses. **Phytotherapy Research**, Vol.21, p.17-25, 2007.

ASARE, G. A.; GYAN, B.; BUGYEI, K.; ADJEI, S.; MAHAMA, R.; ADDO, P.; OTU-NYARKO, L.; WIRED, E. K.; NYARKO, A. Toxicity potentials of the nutraceutical *Moringa oleifera* at supra-supplementation levels. **Journal of Ethnopharmacology**, Vol.139, p.265- 272, 2012.

AWODELE, O.; OREAGBA, I. A.; ODOMA, S.; SILVA, J. A. T.; OSUNKALU, V. O. Toxicological evaluation of the aqueous leaf extract of *Moringa oleifera* Lam. (Moringaceae). **Journal of Ethnopharmacology**, Vol.139, p.330- 336, 2012.

AZMIR, J.; ZAIDUL, I. S. M.; RAHMAN, M. M.; SHARIF, K. M.; MOHAMED A.; SAHENA, F.; JAHURUL, M. H. A.; GHAFOOR, K.; NORULAINI, N.A.N.; OMAR, A.K.M. Techniques for extraction of bioactive compounds from plant materials: A review. **Journal of Food Engineering**. p. 426-436, 2013.

BAXI, S. N.; PORTNOY, J. M.; LARENAS-LINNEMANN, D.; PHIPATANAKUL, W. Exposure and Health Effects of Fungi on Humans. **J Allergy Clin Immunol Pract**, p.396-404, 2016.

BERLINCK, R. G. S.; BORGES, W. S; SCOTTI, M. T.; VIEIRA, P. C. A química de produtos naturais do brasil do século XXI. **Quim. Nova,** v. 40, n. 6, p. 706-710, 2017.

BEZERRA, I. N.; MOREIRA, T. M. V.; CAVALCANTE, J. B.; SOUZA, A. M.; SICHIERI, R. Consumption of food away from home in Brazil according to places of purchase. **Revista de Saùde Pùblica**, 2017.

BRAZIL. Ministry of Health. Secretariat of Science, Technology and Strategic Inputs. Department of Pharmaceutical Assistance. **National policy on medicinal plants and herbal medicines**. Brasilia: Ministry of Health, 2006.

BRILHANTE, R. S. N.; SALES, J. A.; PEREIRA, V. S.; CASTELO-BRANCO, D. S. C. M.; CORDEIRO, R. A.; SAMPAIO, C. M. S.; PAIVA, M. A. N.; SANTOS, J. B. F.SIDRIM, J. J. C.; ROCHA, M. F. G. Research advances on the multiple uses of *Moringa oleifera*: A sustainable alternative for socially neglected population. **Asian Pacific Journal of Tropical Medicine**, p.1-10, 2017.

CELESTINO, S. M. C. **Principios de Secagem de Alimentos**. Planaltina: Embrapa, 2010.

CLARO, R. M.; SANTOS, M. A. S.; OLIVEIRA, T. P.; PEREIRA, C. A.; SZWARCWALD, C. L.; MALTA, D. C. Consumption of unhealthy foods related to chronic non-communicable diseases in Brazil: National Health Survey, 2013. **Epidemiol. Serv. Saùde**. 2015.

COORDINATION FOR THE DEVELOPMENT OF HIGHER EDUCATION STAFF.

Tables of Areas of Knowledge. Available at:

<https://www.capes.gov.br/images/stories/download/avaliacao/TabelaAreasConhecimento_072012.pdf>. Accessed on May 23, 2017.

FERREIRA, P. M. P.; FARIAS, D. F.; OLIVEIRA, J. T. A.; CARVALHO, A. F. U. *Moringa oleifera*: bioactive compounds and nutritional potential. **Rev. Nutr.**, Campinas, v.21, p.431-437, jul./ago., 2008.

FIGUEREDO, C. A.; GURGEL, I. G. D.; JUNIOR, G. D. G. A política nacional de plantas medicinais e fitoterâpicos: construção, perspectivas e desafios. **Revista de Saùde Coletiva**, p. 381-400, 2014.

FILHO, F. E. T. **Natural or synthetic vitamins: learn about the causes of nutritional poverty in food**. VitalBio - Bioenergized natural vitamins, 2017.

FRANCO, B. D. G. M.; LANDGRAF, M. **Microbiologia dos Alimentos**. Sao Paulo: Atheneu, 2004.

FUGLIE, L. J. **The Miracle Tree: *Moringa oleifera*: Natural Nutrition for the Tropics**. Training Manual, 2002.

GASQUI, D. L.; MARINELLI, P. S.; OTOBONI, A. M. M. B.; TANAKA, A. Y.; OLIVEIRA, A. S. **Chemical and nutritional characterization of moringa flour (*Moringa oleifera* Lam.)**. 2013.

GAVA, A. J. **Principios de tecnologia de alimentos**. Sao Paulo: Nobel, 1984.

GIL, A. C. **Como elaborar projetos de pesquisa**. 4. ed. Sao Paulo: Atlas, 2002.

GIL, A. C. **Métodos e técnicas de pesquisa social**. 6. ed. Sao Paulo: Atlas, 2008.

GONÇALVES, E. C. B. A. **Anâlise de alimentos: uma visâo quimica da nutrição**. Sao Paulo: Livraria Varela, 2006.

GOPALAKRISHNAN, L.; DORIYA, K.; KUMAR, D. S. *Moringa Oleifera*: A Review on Nutritive Importance and its Medicinal Application. **Food Science and Human Wellness**, 2016.

GOPALAN, C.; RAMA SASTRI, B. V.; BALASUBRAMANIAN, S. C. **Nutritive value of Indian foods**. Hyderabad, India: (National Institute of Nutrition), 1971.

GUPTA, S.; JAIN, R.; KACHHWAHA, S.; KOTHARI, S. L. Nutritional and medicinal applications of *Moringa Oleifera* Lam. - Review of current status and future possibilities. **Journal of Herbal Medicine**, 2017.

HARVEY, R. A.; FERRIER, D. R. **Illustrated biochemistry**. 5. ed. Porto Alegre : Artmed, 2012.

ADOLFO LUTZ INSTITUTE. **Chemical and physical methods for food analysis**. Sao Paulo, 2008,

KARIM, N. A. A.; IBRAHIM, M. D.; KNTAYYA, S. B.; RUKAYADI, Y.; HAMID, H. A.; RAZIS, A. F. A. *Moringa oleifera* Lam: Targeting Chemoprevention. **Asian Pacific Journal of Cancer Prevention**, Vol.17, 2016.

KAUR, A.; KAUR, P. K.; SINGH, S.; SINGH, I. P. Antileishmanial compounds from *Moringa oleifera* Lam. **Z. Naturforsch**, p.110 - 116, 2014.

LEONE, A.; FIORILLO, G.; CRISCUOLI, F.; RAVASENGHI, S.; SANTAGOSTINI, L.; FICO, G.; SPADAFRANCA, A.; BATTEZZATI, A.; SCHIRALDI, A.; POZZI, F.; LELLO, S.FILIPPINI, S.; BERTOLI, S. Nutritional Characterization and Phenolic Profiling of *Moringa oleifera* Leaves Grown in Chad, Sahrawi Refugee Camps, and Haiti. **International Journal of Molecular Sciences**, Vol.16, p.18923-189372015, 2015.

LEONE, A.; SPADA, A.; BATTEZZATI, A.; SCHIRALDI, A.; ARISTIL, J.; BERTOLI, S. Cultivation, Genetic, Ethnopharmacology, Phytochemistry and Pharmacology of *Moringa oleifera* Leaves: An Overview. **International Journal of Molecular Sciences**, Vol.16, p.12791-12835, 2015.

MAHAN, L. K.; ESCOTT-STUMP, S.; RAYMOND, J. L.; **Krause Food Nutrition and Diet Therapy**. 13ª ed. Elsevier, 2013.

MANSOUR, H. H.; ISMAEL, N. E. R.; HAFEZ, H. F. Modulatory effect of *Moringa oleifera* against gamma-radiation-induced oxidative stress in rats. **Biomedicine & Aging Pathology**, p.265-272, 2014.

MARTiNEZ-GONZALEZ, C. L.; MARTiNEZ, L.; MARTiNEZ-ORTIZ, E. J.; GONZALEZ-TRUJANO, M. E.; DÉCIGA-CAMPOS, M.; VENTURA-MARTiNEZ, R.; DÌAZ-REVAL, I. *Moringa oleifera,* a species with potential analgesic and anti-inflammatory activities. **Biomedicine & Pharmacotherapy**, p.482-488, 2017.

MATHUR, B. S. **Moringa book**. Trees For Life, 2005.

MORAIS, D. C.; MORAES, L. F. S.; SILVA, D. C. G.; PINTO, C. A.; NOVAES, J. F. Methodological aspects of diet quality assessment in Brazil: a systematic review. **Ciência e Saùde Coletiva**, 2016.

MOYO, B.; MASIKA, P. J.; HUGO, A.; MUCHENJE, V. Nutritional characterization of Moringa (*Moringa oleifera* Lam.) leaves. **African Journal of Biotechnology**, Vol.10, p. 12925-12933, October, 2011.

MOYO, B.; OYEDEMI, S.; MASIKA, P. J.; MUCHENJE, V. Polyphenolic content and antioxidant properties of *Moringa oleifera* leaf extracts and enzymatic activity of liver from goats supplemented with *Moringa oleifera* leaves/sunflower seed cake. **Meat Science**, Vol.91, p.441-447, 2012.

NELSON, D. L.; COX, M. M. **Lehninger's principles of biochemistry**. 6. ed. Porto Alegre: Artmed, 2014.

OLSON, M. E.; SANKARAN, R. P.; FAHEY, J. W.; GRUSAK, M. A.; ODEE, D.; NOUMAN, W. Leaf Protein and Mineral Concentrations across the "Miracle Tree" Genus Moringa. **PLoS ONE**, 2016.

OSBORNE, D. P.; VOOGT, P. **Calculation of calorific value**. In: The Analysis of Nutrients in Foods. London: Academic Press, 1978.

OYAGBEMI, A. A.; OMOBOWALE, T. O.; AZEEZ, I. O.; ABIOLA, J. O.; ADEDOKUN, R. A. M.; NOTTIDGE, H. O. Toxicological evaluations of methanolic extract of *Moringa oleifera* leaves in liver and kidney of male Wistar rats. **Journal of Basic and Clinical Physiology and Pharmacology**, Vol.24, p.307-12, 2013.

RANGEL, M. S. A. *Moringa oleifera*: **a plant with multiple uses**. Embrapa: Aracaju, Tabuleiros Costeiros, v.9, mar., 1999.

RAZIS, A. F. A., IBRAHIM, M. D.and KNTAYYA, S. B. Heatlh Benefits of *Moringa oleifera*. **Asian Pacific Journal of Cancer Preventions**, 2014.

SAINI, R. K.; SIVANESAN, I.; KEUM, Y. S. Phytochemicals of *Moringa oleifera*: a review of their nutritional, therapeutic and industrial significance. **3 Biotech**, 2016.

SHIH, M. C.; CHANG, C. M.; KANG, S. M.; TSAI, M. L. Effect of Different Parts (Leaf, Stem and Stalk) and Seasons (Summer and Winter) on the Chemical Compositions and Antioxidant Activity of *Moringa oleifera*. **International Journal of Molecular Sciences**, Vol.12, p.6077-6088, 2011.

SILVA, J. C. et al. **Determination of the chemical composition of the leaves of *Moringa Oleifera Lam*. (moringaceae)**, 2008.

STOHS, S. J.; HARTMAN, M. J. Review of the Safety and Efficacy ofMoringa *oleifera*. **Phytotherapy Research**, Vol.29, p.796-804, 2015.

TEIXEIRA, E. M. B. **Chemical and nutritional characterization of the moringa leaf (*Moringa oleifera Lam*.)**. Universidade Estadual Paulista "Jùlio de Mesquita Filho", Araraquara, 2012.

TEIXEIRA, E. M. B.; CARVALHO, M. R. B.; NEVES, V. A.; SILVA, M. A.; ARANTES-PEREIRA, L. Chemical characteristics and fractionation of proteins from *Moringa oleifera* Lam. leaves. **Food Chemistry**, n. 147, p. 51-54, 2014.

VERGARA-JIMENEZ, M.; ALMATRAFI, M. M.; FERNANDEZ, M. L. Bioactive Components in *Moringa Oleifera* Leaves Protect against Chronic Disease. **Antioxidants**, 2017.

WERNECK, G. L.; HASSELMANN, M. H.; GOUVÊA, T. G. Overview of studies on nutrition and neglected diseases in Brazil. **Ciência e Saùde Coletiva**, p.39-62, 2011.

YAMÉOGO, C. W.; BENGALY, M. D.; SAVADOGO, A.; NIKIEMA, P. A.; TRAORE, S. A. Determination of Chemical Composition and Nutritional Values of *Moringa oleifera* Leaves. **Pakistan Journal of Nutrition**, Vol.10, p.264-268, 2011.

Printed by Books on Demand GmbH, Norderstedt / Germany